Australia • Brazil • Japan • Korea • Mexico • Singapore • Spain • United Kingdom • United States

Bridges

Fast Forward
Yellow Level 7

Text: Alan Trussell-Cullen
Illustrations: Mats Björklund
Editor: Kate McGough
Series design: James Lowe
Design: Vonda Pestana
Production Controller: Emma Hayes
Photo Research: Gillian Cardinal
Reprint: Jennifer Foo

Acknowledgements
The author and publisher would like to acknowledge permission to reproduce material from the following sources: APL/ Corbis/ HJ Martin, p 15 bottom; Getty Images/ PhotoDisc, cover, pp 1, 5 bottom/ Robert Harding World Imagery, p 10 bottom/ AFP, p 15 top; Lonely Planet Images/ Charles Cook, p 9 top; NewsPix, p 14; photolibrary.com/ Carroll Claver, pp 2, 11, 12 top/ Sally Brown, p 4 top/ Hajime Nara, p 5 top/ Jeff Greenberg, p 9 bottom/ Beisch Leigh, p 12 bottom/ Tyler Charles, p 13 bottom; Stock Photos, pp 4 bottom, 6, 8, 13 top.

ISBN 978 0 17 012506 2
ISBN 978 0 17 012511 6 (set)

Cengage Learning Australia
Level 7, 80 Dorcas Street
South Melbourne, Victoria Australia 3205
Phone: 1300 790 853

Cengage Learning New Zealand
Unit 4B Rosedale Office Park
331 Rosedale Road, Albany, North Shore NZ 0632
Phone: 0508 635 766

For learning solutions, visit **cengage.com.au**

Printed in Australia by Ligare Pty Ltd
8 9 10 11 12 13 14 21 20 19 18 17

Evaluated in independent research by staff from the Department of Language, Literacy and Arts Education at the University of Melbourne.

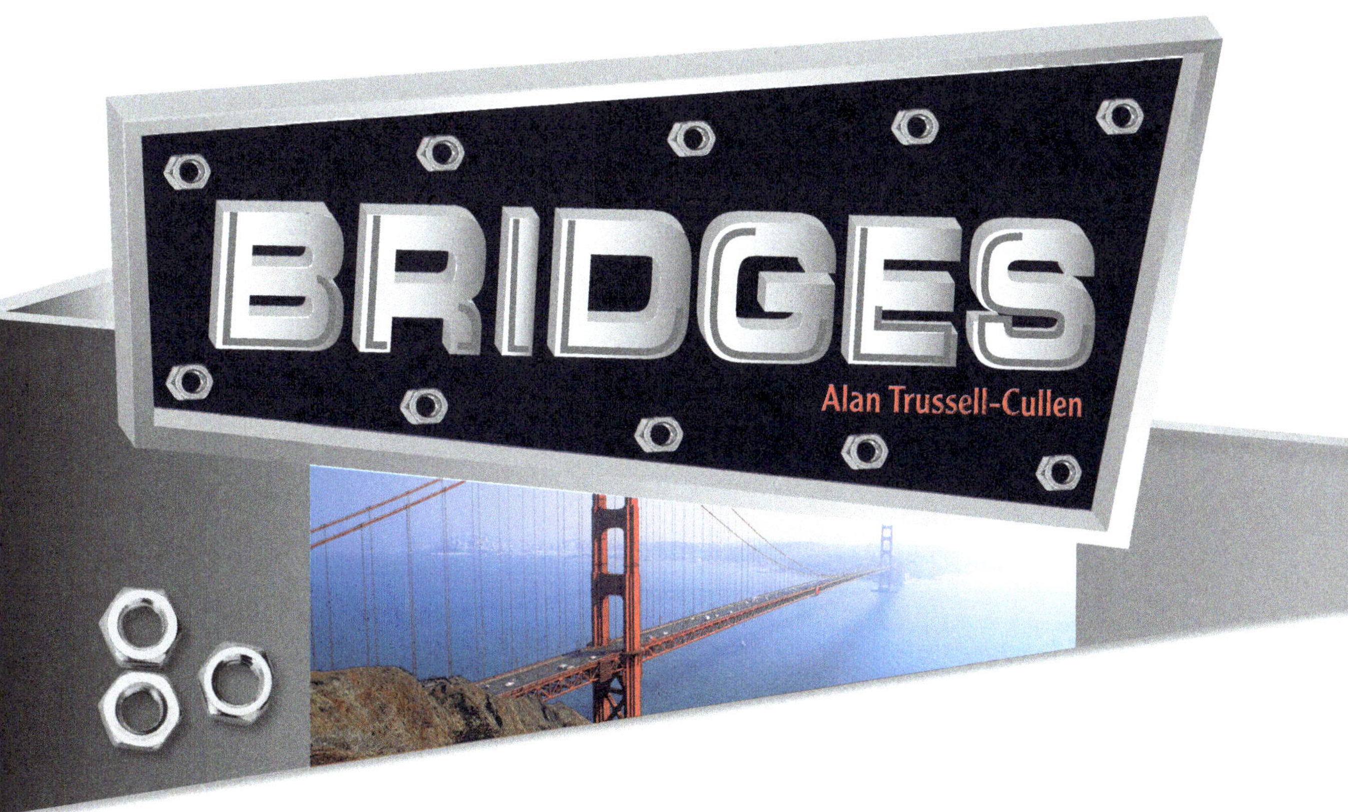

Contents

Chapter 1

BRIDGES

People have been making bridges for thousands of years.

This is a very old bridge.

This is a very new bridge.

Bridges help people cross over water.
Bridges help people cross over land.

Chapter 2

SAFE AND STRONG

There are a lot of cars, buses and people on this bridge.

Bridges must be safe.
They must be very strong to support
all the cars, buses and people.

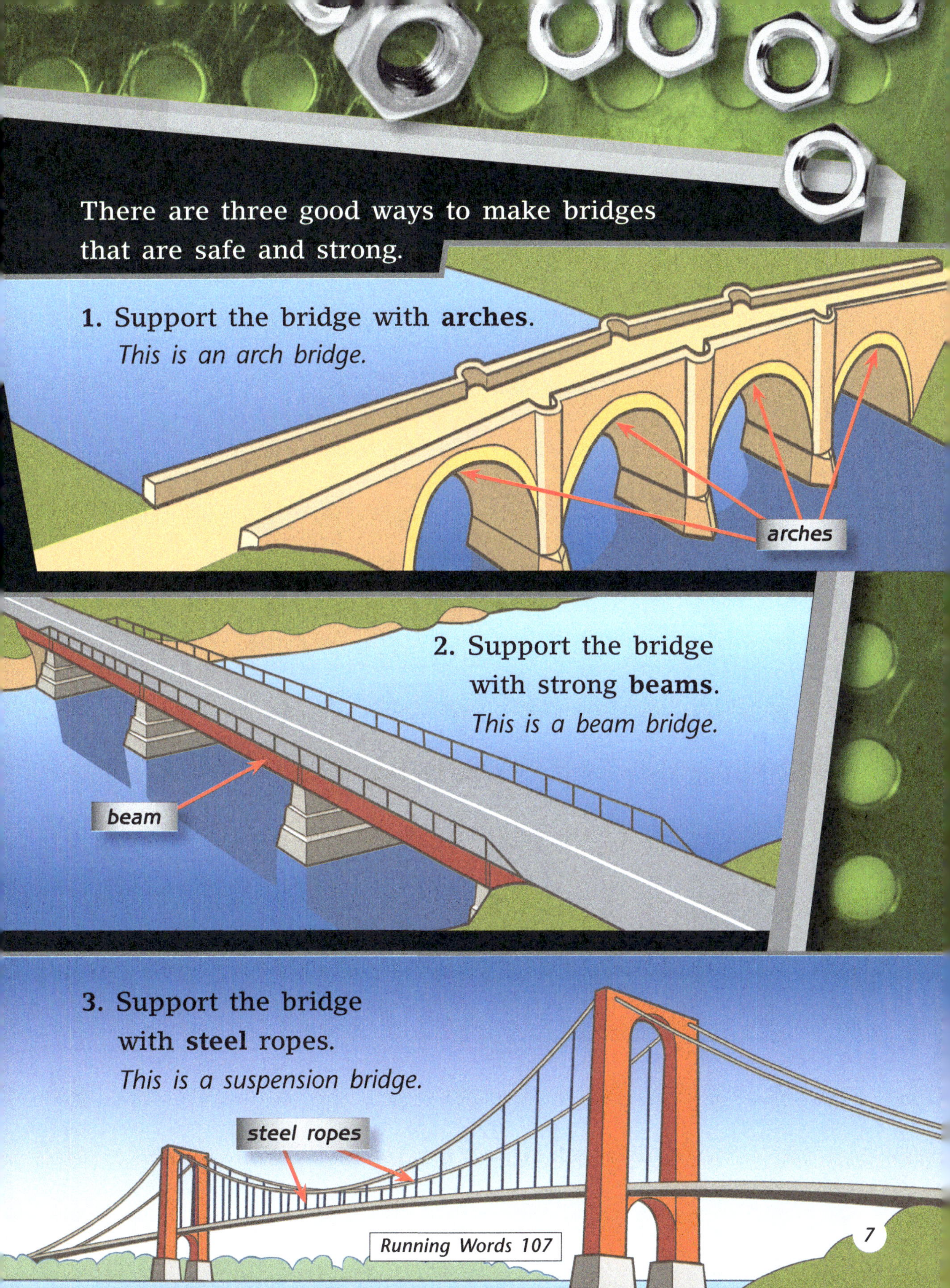

There are three good ways to make bridges that are safe and strong.

1. Support the bridge with **arches**.
   This is an arch bridge.

2. Support the bridge with strong **beams**.
   This is a beam bridge.

3. Support the bridge with **steel** ropes.
   This is a suspension bridge.

Running Words 107

Chapter 3

ARCH BRIDGES

This is what an arch looks like.
An arch is very strong.

This is a very old arch bridge.

Here are some strong arch bridges.

Chapter 4

BEAM BRIDGES

This is a little beam bridge.

This is a big beam bridge.

Some beam bridges have **trusses**. Trusses help make the beams strong. This helps support the bridge.

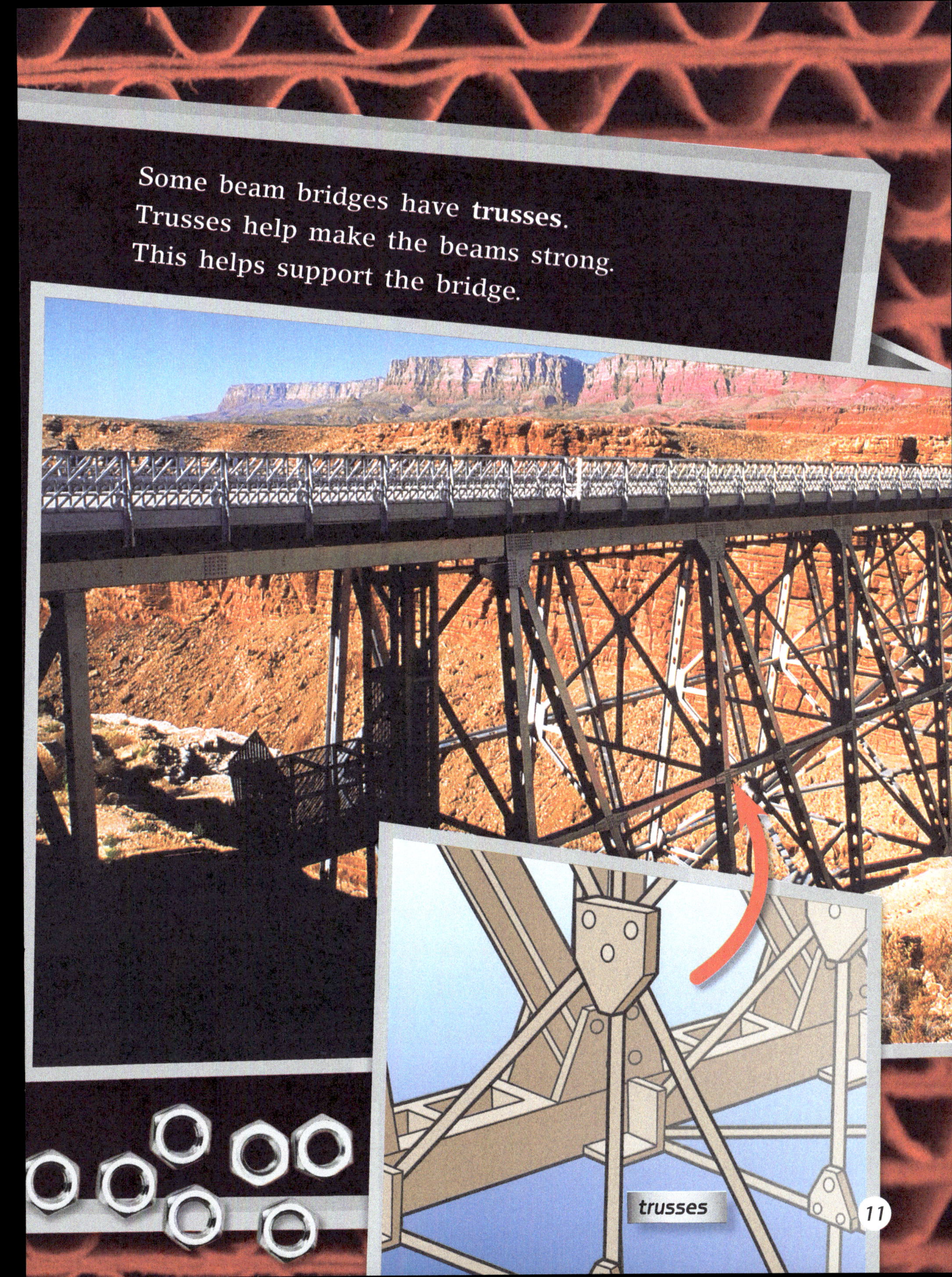

Chapter 5

SUSPENSION BRIDGES

This is a **suspension** bridge.
It has big steel ropes to support it.
The steel ropes have to be big and strong to make the bridge safe.

Here are some suspension bridges.

Chapter 6

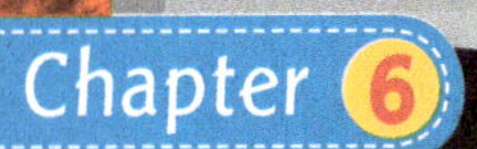

BRIDGE DISASTERS

Most bridges are strong
and do not fall down.
But there have been some bridge **disasters**.

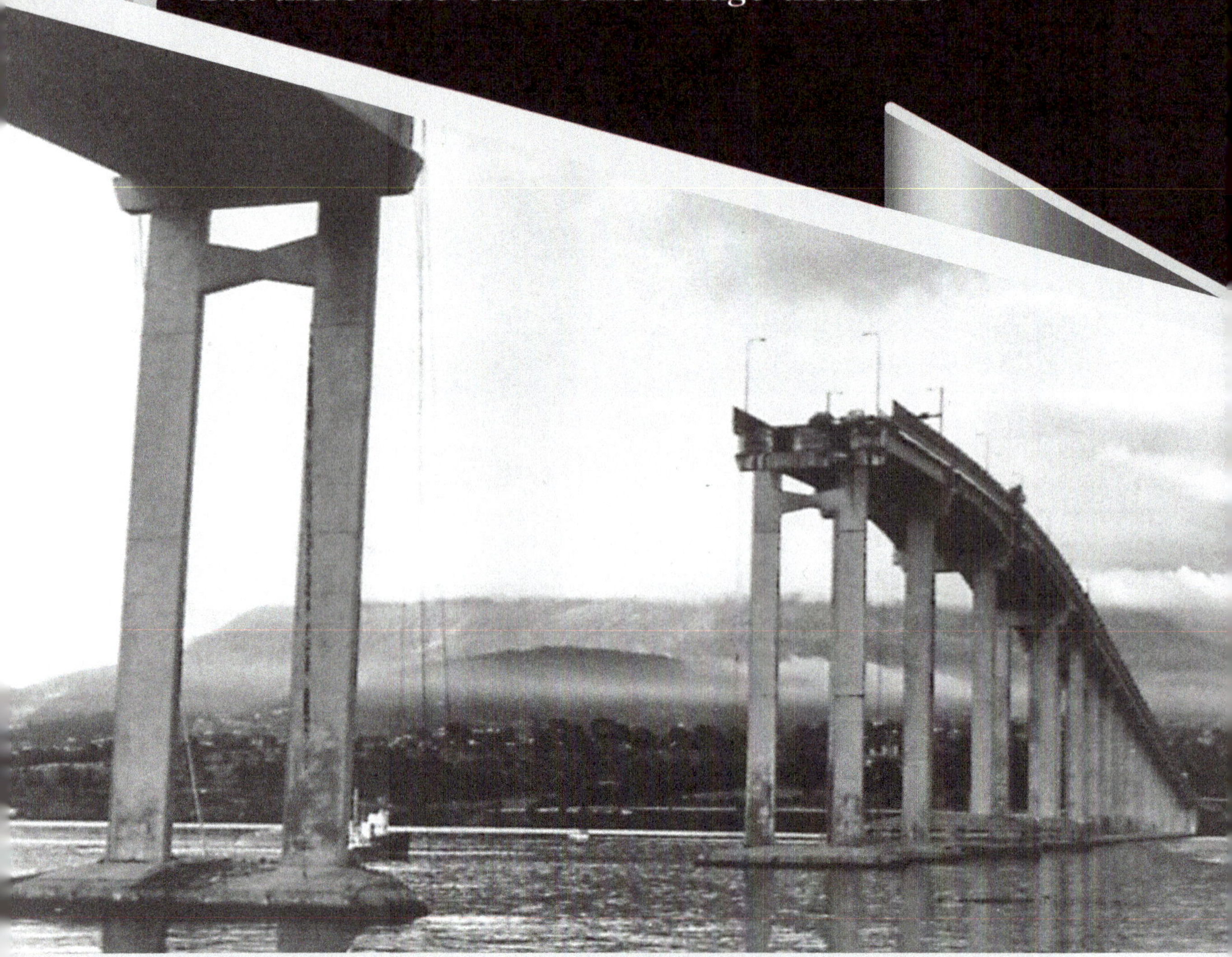

A ship going under this bridge made it fall down.

An **earthquake** made this bridge fall down.

Today, people who make bridges work very hard to make them strong.

Glossary

arches	curves that help hold up bridges
beams	long structures that are part of bridges
disasters	very bad events
earthquake	when the earth moves or breaks open
steel	a type of metal often used when building things
suspension	the act of holding up. Suspension bridges are held up by steel ropes.
trusses	kinds of frames that help hold up bridges

Index